MW00513180

Mars

by Margaret J. Goldstein

Lerner Publications Company • Minneapolis

Lerner Publications Company
A division of Lerner Publishing Group
241 First Avenue North
Minneapolis, MN 55401 USA

Website address: www.lernerbooks.com

Words in **bold type** are explained in a glossary on page 30.

Library of Congress Cataloging-in-Publication Data

Goldstein, Margaret J.
    Mars / by Margaret J. Goldstein.
       p.   cm. — (Our universe)
    Includes index.
    Summary: An introduction to Mars, describing its place in
the solar system, its physical characteristics, its movement
in space, and other facts about this planet.
      ISBN: 0-8225-4651-5 (lib. bdg. : alk. paper)
      1. Mars (Planet)—Juvenile literature. [1. Mars (Planet)]
I. Title. II. Series.
QB641 .G67 2003
523.43—dc21   2002000431

Manufactured in the United States of America
1 2 3 4 5 6 — JR — 08 07 06 05 04 03

The photographs in this book are reproduced with permission from: © USGS/Tsado/Tom Stack & Associates, pp. 3, 15; © John Foster/Photo Researchers, p. 4; © USGS/NASA/Tsado/Tom Stack & Associates p. 8; NASA, pp. 10, 12, 14, 16, 17, 24, 25, 26; © Science VU/Visuals Unlimited, p. 11; Minneapolis Public Library, Planets, pp. 13, 27; © Tsado/NASA/Tom Stack & Associates, p. 19; © Science VU/NASA/Visuals Unlimited, p. 20; © JPL/Tsado/Tom Stack & Associates, p. 21; © Tsado/JPL/NASA/Tom Stack & Associates, p. 23.

Cover: NASA.

This planet has red, rocky ground and a pink sky. Its nickname is the Red Planet. Which planet is it?

It is Mars. Mars is one of our nearest neighbors in space. You can see Mars for yourself at night. It looks like a bright orange star in the sky.

Mars

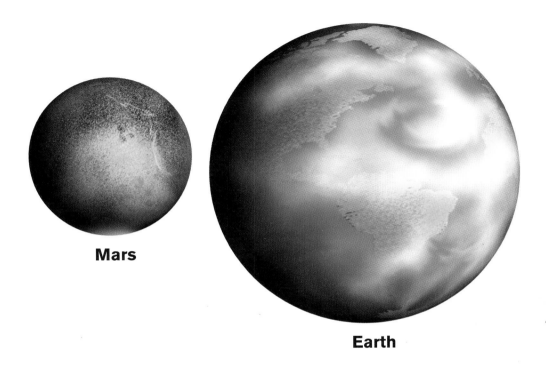

Mars

Earth

Mars is less than half the size of Earth. Mars and Earth are part of the **solar system.** The solar system has nine planets in all.

The nine planets in the solar system **orbit** the Sun. To orbit the Sun is to travel around it. Mars orbits fourth from the Sun. It travels around the Sun between Earth and Jupiter.

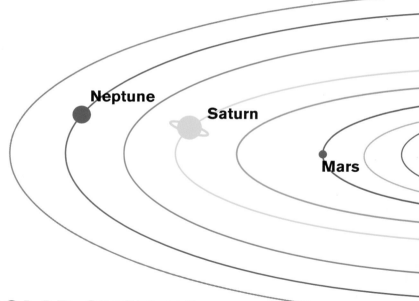

THE SOLAR SYSTEM

Mars takes about two years to make
one full trip around the Sun. Earth
orbits the Sun in one year.

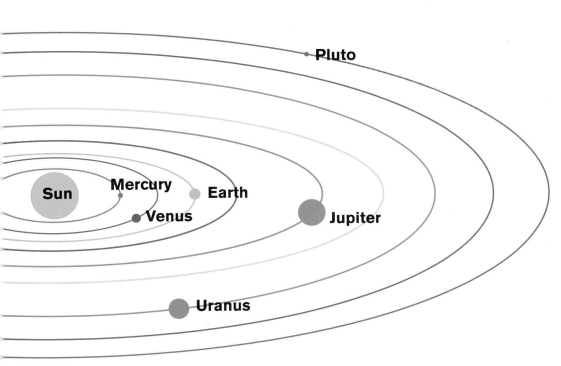

Mars also spins around like a top. This kind of spinning is called **rotating.** Mars rotates all the way around in about 25 hours.

MARS'S LAYERS

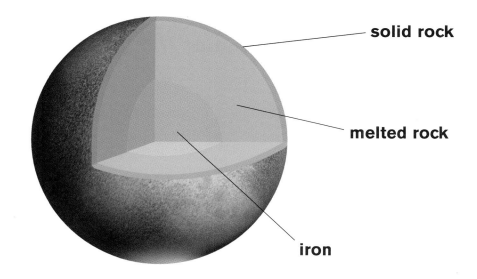

solid rock

melted rock

iron

Mars is a rocky planet. It is covered by an outer layer of rock. A layer of melted rock might lie underneath. The center of Mars might be made of a metal called iron.

Some of the rocks on Mars are big boulders. Others are tiny pieces of sand and dust. Most of the rocks on Mars are red.

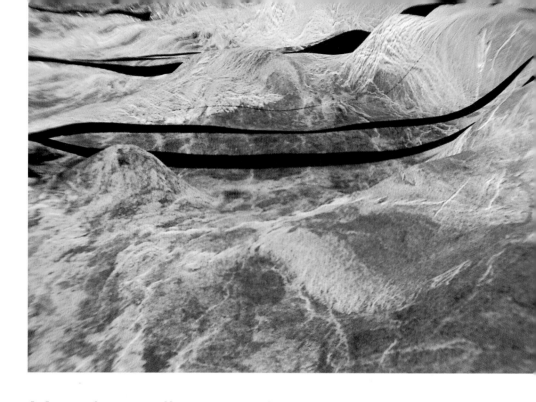

Mars has tall mountains called
volcanoes. They were formed from
hot, melted rock that shot up from
inside the planet. Olympus Mons is the
biggest mountain on Mars.

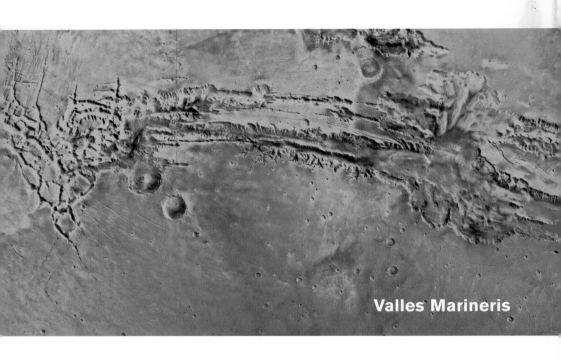

Valles Marineris

Mars has deep valleys called **canyons.**
One very long group of canyons is
called Valles Marineris. On Earth,
Valles Marineris would stretch all the
way across the United States.

12

Mars also has wide pits called **craters.** The craters were made by chunks of rock and metal that crashed into Mars from space.

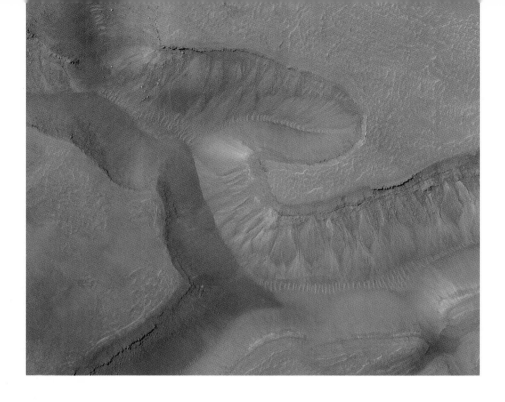

Some places on Mars look like empty
lakes and rivers. These places do not
have any water. But they might have
held water in the past. There might be
some water under the ground on Mars.

Some of the water on Mars is frozen.
Frozen water is called ice. The water is
frozen because Mars is very cold. Mars
gets less warmth from the Sun than
Earth does.

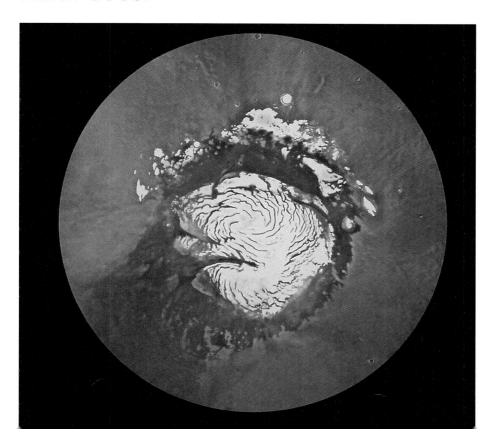

Mars is surrounded by a very thin layer of gases. We call this layer an **atmosphere.** The main gas in Mars's atmosphere is carbon dioxide.

Strong winds blow across the ground on Mars. The winds stir up big clouds of dust. The clouds are red because the dust is red. The red dust clouds make Mars's atmosphere look pink!

Two small moons travel around Mars. Their names are Phobos and Deimos. They look different than Earth's moon.

Earth's moon looks like a round ball. But Mars's moons are not round. What do they look like?

The Moon

Earth

Deimos is shaped like a lumpy potato! It has many small craters. Deimos is one of the smallest moons in the solar system.

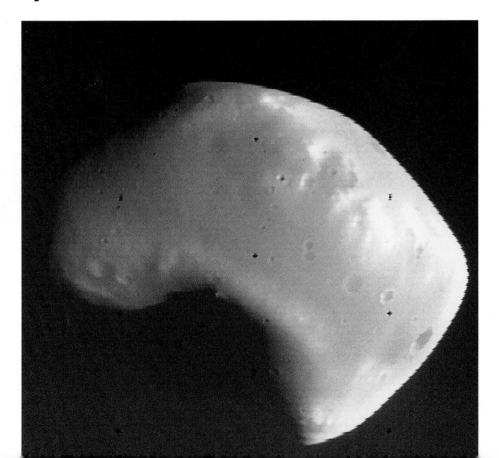

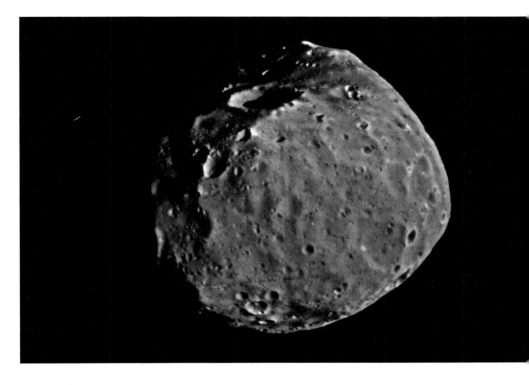

Phobos is about twice as big as
Deimos. It has a big lump on one side
and a big crater on the other. The
crater is named Stickney.

People have wondered about Mars for thousands of years. In 1965, a spacecraft from Earth visited the Red Planet for the first time. The spacecraft was called *Mariner 4.* It took photographs of Mars and studied the planet from space.

Spacecraft called *Viking 1* and *Viking 2* landed on Mars in 1979. They tested the ground, atmosphere, and weather on Mars.

In 1997, a spacecraft called *Pathfinder* landed on Mars. It carried a small car called a rover. The rover tested rocks and dust. It sent information to people on Earth.

People have wondered if any plants or animals live on Mars. But the spacecraft that visited Mars did not find any living things there.

Did Mars once have rivers and lakes full
of water? Have plants or animals ever
lived on the planet? Imagine taking
your own trip to Mars. What questions
would you ask?

Facts about Mars

- Mars is 142,000,000 miles (228,000,000 km) from the Sun.

- Mars's diameter (distance across) is 4,220 miles (6,790 km).

- Mars orbits the Sun in 687 days.

- Mars rotates in 25 hours.

- The average temperature on Mars is −67°F (−55°C).

- Mars's atmosphere is made of carbon dioxide, nitrogen, and argon.

- Mars has 2 moons.

- Mars was named after the Roman god of war.

- The American missions that have visited Mars are the *Mariner* program in 1964–1972, the

Viking program in 1975–1980, *Pathfinder* in 1996–1997, *Global Surveyor* in 1996–present, and *Mars Odyssey* in 2001–present.

- Olympus Mons is twice as high as the tallest mountain on Earth.

- Mars is coldest at its north and south poles. The poles are covered with ice.

- Some clouds in Mars's atmosphere are blue. The blue clouds are made of ice crystals.

- Phobos and Deimos were discovered in 1877.

- Winds on Mars can reach 300 miles (480 km) per hour.

- For a time, people believed that strange creatures lived on Mars. They called the creatures Martians. But there are no Martians.

Glossary

atmosphere: the layer of gases that surrounds a planet or moon

canyon: a deep, narrow valley with steep sides

crater: a large hole on a planet or moon

orbit: to travel around a larger body in space

rotating: spinning around in space

solar system: the Sun and the planets, moons, and other objects that travel around it

volcano: an opening in the surface of a planet where melted rock and gases sometimes burst out

Learn More about Mars

Books

Brimner, Larry Dane. *Mars.* New York: Children's Press, 1999.

Simon, Seymour. *Mars.* New York: Morrow, 1990.

Websites

Solar System Exploration: Mars
<http://solarsystem.nasa.gov/features/planets/mars/mars.html>
Detailed information from the National Aeronautics and Space Administration (NASA) about Mars, with good links to other helpful websites.

The Space Place
<http://spaceplace.jpl.nasa.gov>
An astronomy website for kids developed by NASA's Jet Propulsion Laboratory.

StarChild
<http://starchild.gsfc.nasa.gov/docs/StarChild/StarChild.html>
An online learning center for young astronomers, sponsored by NASA.

Index

atmosphere, 16–17, 24, 28, 29

canyons, 12
craters, 13, 20, 21

Deimos, 18, 20, 29
dust, 17

Earth, 5, 7

life, 26, 29

Mariner spacecraft, 22, 28

Olympus Mons, 11, 29
orbit, 7

Pathfinder spacecraft, 25, 28
Phobos, 18, 21, 29

rocks, 10, 25
rotation, 8, 28

size, 5, 28
solar system, 5–6
Stickney Crater, 21

Valles Marineris, 12
Viking spacecraft, 24, 28
volcanoes, 11

water, 14–15
winds, 17, 29